FUTURE TECHNOLOGY

A GLIMPSE INTO THE FUTURE

CYRIL LAKES

Contents

- CHAPTER ONE 3
- Introduction 3
  - Computing in Quantum 5
  - Future Technology Definition 9
  - New Technologies 13
- CHAPTER TWO 16
  - Genetic engineering and biotechnology 19
  - Sustainable Technologies and Renewable Energy 24
- CHAPTER THREE 31
  - Aerospace Technologies and Space Exploration 31
  - Web of Things (IoT) and Networking 37
  - Modern Manufacturing and Materials 44
  - The Transportation of the Future 51
- CHAPTER FOUR 55
  - Effects on Society and Ethical Issues 59
  - Summary 66
- THE END 70

CHAPTER ONE

Introduction

The exponential growth of technology in the last few decades has sparked conjecture about the revolutionary breakthroughs that could influence our future. The prospects for future technology are both exciting and intimidating, ranging from biotechnology and artificial intelligence to quantum computing and space travel. We will look at some of the most important aspects of future technology in this introduction, along with any possible social repercussions.

Intelligent artificial systems (AI):

Artificial intelligence, or AI, has the potential to completely transform almost every facet of human existence, including banking, entertainment, healthcare, and transportation. AI algorithms will influence our economy, society, and daily lives more and more as they grow more complex and capable of making decisions on their own.

Genetic engineering and biotechnology:

Genetic engineering and biotechnology developments could lead to improved human performance, disease cures, and potentially longer lifespans. Unprecedented control over genetic code is made possible by technologies like CRISPR gene editing, which raises ethical concerns about the possibility of designer babies and genetically altered creatures.

Exploration and Settlement of Space:

With plans for manned flights to Mars, asteroid mining operations, and the creation of permanent colonies on the Moon and beyond, the future of space exploration seems to be revolutionary. Commercial spaceflight enterprises are spearheading rocket technology innovation, increasing private citizen access to space and creating new avenues for resource extraction and exploration.

Computing in Quantum

With exponentially more processing capacity than traditional computers, quantum computing is a paradigm change in computing technology. While still in its infancy, quantum computing has the potential to usher in a new era of creativity and discovery by revolutionizing fields like drug development, materials research, and encryption.

Sustainability and Renewable Energy:

To mitigate climate change and ensure a sustainable future, we must shift to renewable energy sources like solar, wind, and nuclear power. In order to increase the deployment of renewable energy sources and decrease reliance on fossil fuels, significant advancements in energy storage, grid technology, and smart infrastructure are essential.

Virtual reality (VR) and augmented reality (AR):

Technologies such as augmented reality and virtual reality provide immersive online experiences that combine the virtual and real worlds. AR and VR have the power to completely change how we interact with information and the environment around us, from gaming and entertainment to education and training.

IoT, or the Internet of Things:

The network of linked sensors and gadgets that gather and share data in real time is known as the "Internet of Things." As IoT technology develops further, it will make it possible to create intelligent infrastructure systems that improve efficiency, convenience, and safety as well as autonomous cars and smart homes.

Implications for Society and Ethics:

As we adopt these new technology, we also need to consider the ethical and social ramifications. Concerns like algorithmic bias, employment displacement, data privacy, and the digital divide need to be carefully considered and regulated in order to make sure that technology promotes fair outcomes for everyone and serves the greater good.

Technology has a great deal of potential to advance human civilization and solve some of our most important problems in the future. It also brings up difficult moral conundrums and unknowns, though, which need to be handled carefully and sensibly. Through a commitment to innovation, collaboration, and ethical ideals, we can leverage technology to build a more promising and sustainable future for future generations.

Future Technology Definition

The term "future technology" describes scientific, technical, and technological developments, inventions, and breakthroughs that are expected to have a significant impact on society in the ensuing years and decades. These technologies have the power to completely transform a number of facets of human existence, society, and the economy. They frequently mark substantial deviations from existing capabilities or create brand-new paradigms.

Future technology spans many different domains and specialties, such as but not restricted to:

Information technology refers to developments in communications, cybersecurity, data analytics, artificial intelligence, and computers.

Biotechnology: Advances in biomanufacturing, regenerative medicine, genetic engineering, and customized medicine.

Nanotechnology: Advances in nanofabrication methods, nanoelectronics, nanomaterials, and nanomedicine.

Innovations in nuclear fusion, solar energy, wind energy, battery storage, and other sustainable energy sources are referred to as renewable energy.

Space exploration includes advancements in satellite technology, asteroid mining, space flight, and planet colonization.

Quantum computing refers to developments in quantum information science, quantum cryptography, quantum algorithms, and quantum computer hardware.

Automation and robotics: advancements in machine learning, autonomous systems, robotics, and human-machine interfaces.

Virtual reality (VR) and augmented reality (AR) are examples of innovative immersive technologies, along with mixed reality (MR) and extended reality (XR).

The proliferation of smart sensors, cloud computing, edge computing, and Internet of Things (IoT) platforms is known as the Internet of Things (IoT).

Blockchain and Cryptocurrency: Fintech solutions, digital currencies, blockchain applications, and decentralized technology are evolving.

Innovation in engineering, scientific research, government funding, and business endeavors are the engines of future technology. It can solve societal issues, raise living standards, spur economic expansion, and pave the way for new scientific and exploratory endeavors. But it also brings up moral, legal, and societal issues that need to be carefully thought out and managed.

The planet and our collective future will be significantly shaped by technology in the future as long as we keep pushing the boundaries of innovation.

New Technologies

Emerging technologies are breakthroughs and developments that are still in the early phases of development but have the potential to have a big impact on daily life, society, and a number of industries in the near future. These technologies frequently stand for innovative methods, groundbreaking concepts, or fresh uses of previously developed information. The following are a few instances of new technologies:

Artificial intelligence (AI) is the study and application of computer systems that can carry out tasks like speech recognition, decision-making, and problem-solving that normally require human intelligence. Neural networks, deep learning, and machine learning are important branches of AI that are still developing quickly.

Genetic engineering and biotechnology: Biotechnology advances allow researchers to work at the molecular level with biological systems, resulting in novel applications including synthetic biology, tailored medicine, gene editing (e.g., CRISPR-Cas9), and biomanufacturing.

Utilizing the ideas of quantum mechanics, quantum computing is able to process calculations tenfold quicker than traditional computing. Quantum computing holds the potential to transform domains like materials science, medication development, optimization, and cryptography.

Distributed ledger technology, or DLT, and blockchain: DLT and blockchain technology allow for decentralized, transparent, and safe data storage and transactions. Blockchain is used in voting systems, digital identity, supply chain management, and decentralized finance (DeFi), in addition to cryptocurrencies like Bitcoin.

CHAPTER TWO

The term "Internet of Things" (IoT) describes a network of linked sensors and gadgets that gather and share data in real-time. Smart cities, linked cars, industrial automation, precision agriculture, and wearable health monitoring are some of the emerging uses of IoT.

Virtual reality (VR) and augmented reality (AR) technologies combine the virtual and real worlds to produce immersive digital experiences. Virtual training simulations, immersive entertainment, remote collaboration, and interactive marketing are examples of emerging applications.

5G and Beyond: The introduction of 5G networks is expected to bring about improvements in latency, data rates, and connection, opening the door to new developments like remote surgery, augmented reality streaming, self-driving cars, and smart infrastructure.

Automation and Robotics: As robotics technologies develop, they find use in a variety of sectors, including industry, healthcare, agriculture, logistics, and services. The future of work is being transformed by powerful AI-driven robots, autonomous drones, and collaborative robots, or cobots.

Biomedical Technologies: Advances in biomedical engineering, including organ transplantation, tissue engineering, bio-printing, and medical implants, are expanding human potential and transforming healthcare.

Sustainable development and clean energy: New technologies in energy storage, carbon capture, renewable energy, and sustainable materials are tackling environmental issues and accelerating the shift to a low-carbon economy.

These are only a few instances of the wide spectrum of cutting-edge technologies that in the years to come have the power to revolutionize whole industries, spur economic expansion, and enhance quality of life. Businesses, governments, and society at large will face both new opportunities and difficulties as these technologies develop and grow.

Genetic engineering and biotechnology

Science and technology are developing quickly in the disciplines of biotechnology and genetic engineering, which have significant effects on industry, agriculture, the environment, and medicine. An outline of genetic engineering and biotechnology is provided below:

Biotechnology: Using biological systems, organisms, or their derivatives to create goods or applications for particular uses is known as biotechnology. It includes a broad range of approaches and procedures, including as fermentation, molecular biology, genetic engineering, and bioinformatics.

Genetic Engineering: The intentional alteration of an organism's genetic material through the application of biotechnological techniques is a subset of genetic engineering. This can involve modifying, deleting, or adding new genes to produce desired features or attributes.

Uses in the Medical Field:

Gene Therapy: Gene therapies for cancer, genetic abnormalities, and other diseases are being developed using genetic engineering. This is the process of correcting genetic abnormalities or modifying cellular functioning in a patient's cells by adding therapeutic genes.

Biopharmaceuticals: The creation of biopharmaceuticals, which include insulin, vaccinations, monoclonal antibodies, and recombinant proteins made from genetically modified organisms or cell lines, has completely changed the pharmaceutical industry.

Personalized Medicine: Treatments are now able to be customized based on a patient's genetic composition, lifestyle, and health attributes because to developments in genomics and biotechnology.

Uses in the Field of Agriculture:

Genetically Modified Organisms (GMOs): Through genetic engineering, enhanced features like pest resistance, drought tolerance, and higher nutritional content can be developed in crops. GMOs have the ability to improve crop yields, lower the need for pesticides, and solve issues with food security.

Precision Breeding: To speed up the breeding of livestock and crop plants with desired features, biotechnology is also utilized in precision breeding techniques including genome editing and marker-assisted selection.

Industrial Uses:

Biofuel Production: Using renewable resources like agricultural crops, algae, and waste biomass, biotechnology is utilized to create biofuels like ethanol, biodiesel, and biogas.

Bioremediation: The process of employing biological agents to clean up pollution and environmental toxins is known as bioremediation, and it makes use of genetically modified microbes.

Regulatory and Ethical Considerations:

Concerns about safety, environmental effects, and the possibility of unexpected repercussions arise when genetic engineering is used. To guarantee the safety and effectiveness of genetically altered products, regulatory frameworks control their research, testing, and commercialization.

Other topics of discussion in ethics include human gene editing, genetically modified organisms in agriculture, and the possibility of genetic discrimination.

Innovation is still fueled by biotechnology and genetic engineering, which also present promising answers to some of the most important problems confronting society. To ensure the appropriate and sustainable deployment of these technologies, however, rigorous examination of the implications for society, the environment, and ethics is necessary.

Sustainable Technologies and Renewable Energy

The utilization of sustainable technology and renewable energy is vital in tackling worldwide issues including energy security, environmental deterioration, and climate change. By utilizing natural resources that replenish as quickly as they are consumed, these technologies minimize greenhouse gas emissions and lessen dependency on fossil fuels. An outline of sustainable technologies and renewable energy is provided below:

Solar Power:

Solar photovoltaic (PV) technology uses semiconductors like silicon to directly turn sunlight into power.

Through the use of mirrors or lenses, solar thermal technology concentrates sunlight to produce heat that can be utilized to heat buildings or create energy.

Solar energy has the capacity to supply a sizable amount of the world's electricity needs because it is plentiful, clean, and renewable.

Wind Power:

Wind energy is captured by wind turbines and transformed into mechanical power, which generators use to create electricity.

Wind energy is captured by onshore and offshore wind farms, which produce electricity at utility size and add to the global renewable energy capacity.

Hydropower:

Using turbines and dams, hydropower converts the energy of flowing water into electrical power.

One of the earliest and most popular renewable energy sources, hydropower offers a dependable and dispatchable supply of electricity in many areas.

Energy from Biomass:

Biomass energy can be used to generate heat, electricity, or biofuels from organic resources like wood, waste biomass, and agricultural wastes.

Anaerobic digestion, gasification, combustion, and the processes used to produce biofuels are examples of bioenergy technology.

Geothermal Power:

Geothermal energy uses heat that is trapped beneath the surface of the Earth to produce electricity or heat.

Geothermal power stations use hot water or steam from underground reservoirs to turn turbines and produce energy.

Integrated and Hybrid Systems:

In order to maximize energy production and increase dependability, hybrid renewable energy systems integrate a number of renewable energy sources, including solar, wind, and storage technologies.

Demand-side management, energy storage, smart grid technology, and renewable energy are all combined in integrated energy systems to produce more sustainable and resilient energy infrastructure.

Energy Retention:

Variable renewable energy sources can be integrated into the grid thanks in large part to energy storage technologies including batteries, pumped hydro storage, and thermal energy storage.

Large-scale renewable energy deployment is being fueled by improvements in energy storage efficiency, cost, and capacity, which also promote grid stability and dependability.

Energy management and the smart grid:

Demand response and grid balancing services are made possible by smart grid technologies, which also maximize the integration of renewable energy sources and increase grid efficiency.

Digital platforms and energy management systems offer real-time monitoring, control, and optimization of energy consumption, promoting sustainability and energy efficiency.

The shift towards a low-carbon economy and a more robust and equitable energy system is contingent upon the utilization of sustainable technology and renewable energy. A sustainable future for future generations and the attainment of global climate targets depend on investments made in research, development, and implementation of these technologies.

CHAPTER THREE

Aerospace Technologies and Space Exploration

Aerospace technologies and space exploration have revolutionized our understanding of the cosmos and created new avenues for technological advancement, scientific research, and human exploration. An outline of aircraft technologies and space exploration is provided below:

Launchers:

Spacecraft are propelled into orbit and beyond Earth's atmosphere by rockets. These vehicles range from heavy-lift launchers that can carry people and payloads to the Moon, Mars, and beyond, to tiny, disposable rockets for satellite placement.

Satellites:

Artificial objects known as satellites are positioned in orbit around the Earth or other celestial bodies for various purposes such as communication, navigation, Earth observation, and scientific study. There are several varieties of these satellites, such as low Earth orbit, polar orbit, and geostationary.

Rovers and Probes in Space:

Unmanned spacecraft called space probes and rovers are made to investigate far-off planets, moons, asteroids, and comets. NASA's Mars rovers, such as Curiosity and Perseverance, which explore the Martian surface, and the Voyager spacecraft, which have traveled into interstellar space, are two examples.

Human Space Flight:

Sending humans into space to carry out research, maintain space stations and spaceships, and explore celestial bodies is known as human spaceflight. The International Space Station (ISS) is a cooperative research platform used for conducting in-orbit experiments and investigating the effects of microgravity on the human body.

Space-Based Telescopes:

Space telescopes are observatories launched into orbit to study celestial objects and processes without the atmospheric distortions caused by Earth. Two prime examples are the James Webb Space Telescope, which is scheduled for flight in the near future, and the Hubble Space Telescope, which has transformed our understanding of the cosmos.

Travel to Space:

The goal of commercial space tourism enterprises is to open up space travel to the general public. Companies like SpaceX, Blue Origin, and Virgin Galactic are offering experiences like lunar travel, suborbital spaceflights, and orbital space vacations.

Investigating Planets:

Missions dedicated to planetary exploration investigate planets, moons, and other celestial bodies both inside and outside of our solar system. These missions shed light on the possible habitability, atmosphere, and geology of the planets. Examples are the ESA's Rosetta mission to comet 67P/Churyumov-Gerasimenko and NASA's Mars Exploration Program.

Propulsion and Materials in Aerospace:

The development of lightweight composites and heat-resistant alloys, among other aerospace materials, has made it possible to build spacecraft and airplanes that can endure harsh environments both in space and in Earth's atmosphere.

Propulsion technologies such as nuclear, ion, and chemical rockets are always developing to increase thrust, efficiency, and travel possibilities in space exploration.

Space Settlement and Colonization:

Permanent human colonies on the Moon, Mars, and other celestial worlds are among the long-term goals of space exploration. Developments in habitat design, life support systems, and sustainable resource use in space are needed for these initiatives.

Aerospace technologies and space exploration continue to push the limits of human knowledge and capability, fostering worldwide cooperation, scientific breakthroughs, and technological innovation. By advancing science, technology, and exploration, these space missions have the potential to aid mankind on Earth while also addressing fundamental issues regarding the universe's beginnings and the possibility of life beyond Earth.

Web of Things (IoT) and Networking

A network of networked sensors, actuators, gadgets, and other items that gather and share data online is known as the Internet of Things (IoT). These devices can communicate with central systems and with one other, frequently without the need for human intervention, because they are embedded with electronics, software, and network connectivity. An outline of connectivity and IoT is provided below:

Connectivity of Devices:

Many wireless and wired communication methods, such as satellite, cellular networks (e.g., 4G LTE, 5G), Bluetooth, Wi-Fi, and low-power wide-area networks (LPWANs) including LoRaWAN and NB-IoT, are used by IoT devices to connect to the internet.

Actuators and Sensors:

Sensors on Internet of Things devices track and detect ambient factors like light, motion, sound, humidity, temperature, and pressure. Based on data inputs, actuators allow devices to carry out operations or regulate physical systems.

Data Processing and Transmission:

IoT devices gather sensor data and send it for processing and analysis to edge computing nodes, cloud platforms, or centralized servers. Decision-making and actionable insights are made possible by machine learning algorithms and real-time data analytics.

Use cases and applications:

The Internet of Things (IoT) has applications in many different businesses and fields, such as manufacturing, energy, smart cities, smart homes, smart healthcare, smart agriculture, smart transportation, and environmental monitoring.

Smart thermostats, wearable health monitors, linked cars, industrial automation, precision agricultural systems, and intelligent infrastructure are a few examples of IoT application cases.

Collaboration and Commons:

For seamless communication and data sharing, different IoT devices and platforms must be compatible with one another. Interoperability is aided and facilitated by standardization initiatives undertaken by groups like industry consortia, the Internet Engineering Task Force, and the Institute of Electrical and Electronics Engineers (IEEE).

IoT messaging and communication typically use protocols like MQTT, CoAP, HTTP, and AMQP.

Privacy and Security:

Because there are so many linked devices and possible security holes, security is a top priority in Internet of Things implementations. Intrusion detection systems, data encryption, secure bootstrapping, over-the-air (OTA) updates, and device authentication are examples of IoT security methods.

Protecting personal data gathered by Internet of Things (IoT) devices and guaranteeing adherence to data protection laws like the California Consumer Privacy Act (CCPA) and the General Data Protection Regulation (GDPR) are two privacy considerations.

Computing on the Edge and Fog:

By distributing processing power and data processing closer to Internet of Things (IoT) devices, edge computing and fog computing systems lower latency, consumption of bandwidth, and dependency on centralized cloud infrastructure. Before sending pertinent data to the cloud, edge devices and gateways do data preprocessing, filtering, and analysis at the network edge.

Prospective Patterns and Obstacles:

The expansion of 5G networks, the incorporation of AI and ML algorithms into IoT systems, the use of blockchain for safe data sharing and transactions, and the creation of autonomous and decentralized IoT ecosystems are some of the anticipated developments in the field of IoT.

For IoT research, development, and deployment, issues like scalability, interoperability, security flaws, data privacy concerns, regulatory compliance, and energy efficiency continue to be priority topics.

Connectivity technologies and the Internet of Things (IoT) are driving innovation, digital transformation, and the development of new services and business models in a variety of industries. IoT has the ability to address global concerns and create new opportunities for economic growth and sustainable development while also enhancing efficiency, productivity, and quality of life. These advancements in connectivity, computation, and data analytics make IoT a promising technology.

Modern Manufacturing and Materials

Technological advancements in materials science and manufacturing are transforming entire industries by facilitating the creation of new goods with superior qualities, more efficiency, and less environmental effect. An outline of advanced materials and production is provided below:

Advanced Substances:

Engineered materials with superior qualities or functions over conventional materials are known as advanced materials. These materials frequently have special mechanical, thermal, electrical, or optical properties that allow for creative applications in a range of sectors.

Composites, ceramics, polymers, metals and alloys, semiconductors, nanomaterials, biomaterials, and smart materials (such as shape-memory alloys and self-healing polymers) are a few examples of advanced materials.

Nanotechnology:

By modifying matter at the nanoscale, which is often between 1 and 100 nanometers, nanotechnology is able to produce materials and devices with unique qualities and capabilities. Nanomaterials' high surface-to-volume ratio and quantum processes give rise to special size-dependent characteristics.

The fields of nanoelectronics, nanomedicine, nanocomposites, nanophotonics, nanomagnetics, and nanosensors are among the applications of nanotechnology.

3D Printing and Additive Manufacturing:

Additive manufacturing, commonly referred to as 3D printing, makes it possible to construct intricate three-dimensional objects directly from digital plans, layer by layer. With traditional production processes, customisation, rapid prototyping, and design flexibility are not achievable. These features are provided by this technology.

Fused deposition modeling (FDM), electron beam melting (EBM), stereolithography (SLA), selective laser sintering (SLS), and direct metal laser sintering (DMLS) are examples of additive manufacturing methods.

Modern Manufacturing Techniques:

Advanced manufacturing refers to a broad range of procedures and methods used to effectively and economically produce high-quality parts and products. Precision machining, laser cutting and welding, injection molding, casting, forging, and extrusion are some of these procedures.

Cutting-edge manufacturing methods facilitate the creation of complicated geometries and useful surfaces while increasing productivity, decreasing waste, and optimizing resource usage.

Intelligent Production and Industry 4.0:

Smart manufacturing creates intelligent, networked factories by incorporating automation, digital technologies, and data analytics into production processes. Industry 4.0 projects use robotics, cloud computing, artificial intelligence (AI), Internet of Things (IoT) devices, and cyber-physical systems to increase manufacturing operations' responsiveness, flexibility, and efficiency.

Supply chain optimization, quality assurance, predictive maintenance, real-time monitoring, and flexible production in response to shifting consumer demands are all made possible by smart manufacturing.

Surface treatments and advanced coatings:

The performance, robustness, and utility of materials and components are improved by sophisticated coatings and surface treatments. These coatings offer anti-fouling qualities, thermal insulation, wear resistance, corrosion resistance, and beautiful finishes.

Nanocoatings, thin films, self-cleaning coatings, thermal barrier coatings, and functionalized surfaces for electrical and medicinal applications are a few examples of advanced coatings.

Sustainability of the Environment:

Modern manufacturing techniques and materials use less energy, produce less waste, and make better use of their resources, all of which support environmental sustainability. Eco-friendly materials, energy-efficient procedures, material recycling and reuse, and pollution control are the main focuses of sustainable manufacturing methods.

New Developments and Prospects:

Bioinspired materials, metamaterials, 4D printing (materials that change shape over time), sustainable materials and manufacturing techniques, and the incorporation of AI and machine learning for autonomous optimization and decision-making in manufacturing systems are some of the emerging trends in advanced materials and manufacturing.

Technological advancements in manufacturing and advanced materials continue to fuel economic growth, industry innovation, and competitiveness. These technologies hold the ability to address global concerns, enhance quality of life, and produce sustainable solutions for the future by facilitating the creation of novel materials, products, and processes.

The Transportation of the Future

Urbanization, environmental concerns, technological advancements, and shifting consumer preferences are all contributing to a significant alteration of the transportation landscape in the coming years. Below is a summary of some of the major themes and advancements influencing transportation in the future:

Electric-powered cars (EVs):

The automobile industry is undergoing a transformation as a result of the growing usage of electric vehicles, which also reduces greenhouse gas emissions and dependency on fossil fuels. Globally, EV sales are increasing due to advancements in battery technology, charging infrastructure, and government incentives.

Self-driving cars (AVs):

Self-driving automobiles, also known as autonomous vehicles, have the potential to completely transform transportation by increasing accessibility, efficiency, and safety. In order to navigate roadways and make decisions about driving in real time, autonomous vehicles (AVs) rely on sensors, cameras, radar, lidar, and artificial intelligence algorithms.

Ride-Hailing and Shared Mobility:

The proliferation of shared mobility services, including ride-hailing, car-sharing, and micro-mobility (such as electric scooters and bikes), is changing the face of urban transportation and diminishing the necessity of owning a private vehicle. Platforms for shared mobility provide affordable, practical substitutes for conventional sources of transportation.

The Hyperloop and High-Speed Rail:

With their quick, sustainable, and energy-efficient modes of transportation, high-speed rail and hyperloop technologies have the potential to completely transform long-distance travel. By enabling quick connections between cities and regions, these transit options cut down on travel times and airport and road traffic.

UAM, or urban air mobility:

Urban air mobility refers to the on-demand air transportation within urban areas made possible by drones, flying taxis, and electric vertical takeoff and landing (eVTOL) aircraft.

CHAPTER FOUR

UAM technologies offer effective airborne passenger and cargo transportation as well as potential solutions to urban traffic.

Intelligent Transportation Systems Infrastructure:

In order to maximize traffic flow, lessen congestion, and enhance safety, smart transportation infrastructure incorporates digital technologies, sensors, and data analytics into public transportation systems, roadways, and traffic lights. Adaptive traffic management, predictive analytics, and real-time monitoring are made possible by intelligent transportation systems (ITS).

Ecological Air Travel:

To lessen its environmental impact and become carbon neutral, the aviation sector is investing in electric and hybrid aircraft, carbon offsetting programs, and sustainable aviation fuels (SAFs). The goal of sustainable aviation technology is to reduce commercial aviation's fuel consumption, pollutants, and noise pollution.

Intermodal Communication:

Intermodal transportation networks offer convenient door-to-door mobility options for both passengers and freight by smoothly integrating several forms of transportation, including buses, trains, bicycles, and electric vehicles. Integration of ticketing systems and intermodal hubs enable seamless transitions between transportation modes.

Policy and Regulatory Initiatives:

Governments everywhere are putting laws, subsidies, and policies into place to encourage the use of electric vehicles, enhance public transportation systems, and lower emissions from the transportation industry. Transportation policy decisions are influenced by urban planning techniques, air quality regulations, and climate targets.

Connectivity and Digital Transformation:

The transportation ecosystem is changing as a result of digitalization and connectivity, which make it possible for real-time communication, navigation, payment transactions, and customized travel experiences. Digital marketplaces, data-sharing platforms, and mobile apps improve the efficiency, accessibility, and convenience of transportation services.

Innovation, sustainability, and integration will define the transportation of the future, with an emphasis on lowering environmental effect, expanding mobility options, and raising systemic efficiency. The transportation industry is in a good position to construct more robust and sustainable transportation networks for the future while meeting the changing demands and preferences of travelers by utilizing cutting-edge technologies, data-driven insights, and cooperative collaborations.

Effects on Society and Ethical Issues

To ensure fair, sustainable, and socially responsible solutions, a number of societal implications and ethical issues related to the future of transportation must be carefully considered. Some major areas of concern are as follows:

Fair Access

Ensuring fair access to transportation options for all members of society is crucial when transportation systems undergo evolution. This entails tackling concerns related to accessibility for those with impairments, affordability, and transportation deserts in impoverished areas.

Effect on the Environment:

A major factor in air pollution, greenhouse gas emissions, and climate change is transportation. Making the switch to greener, more sustainable forms of transportation, like public transportation and electric cars, is essential to minimizing environmental damage and the effects of climate change.

Security and Safety:

Passengers, pedestrians, and other road users must all be safe and secure due to the advent of autonomous vehicles and urban air transportation. Establishing liability frameworks, addressing cybersecurity risks, and reducing the likelihood of accidents and collisions are a few examples of ethical considerations.

Security and Privacy of Data:

Large volumes of data regarding the movements, actions, and preferences of passengers are produced by connected and autonomous cars. The development and application of transportation technology must take ethical issues into account in order to safeguard individuals' right to privacy and prevent sensitive data from being misused or accessed without authorization.

Loss of Employment:

Traditional employment patterns could be disrupted by the broad deployment of automation and autonomous technology in transportation, especially in sectors like ridesharing, delivery services, and trucking. Addressing the effect on workers' livelihoods, offering aid with retraining and transition, and guaranteeing fair labor standards are all examples of ethical considerations.

Urbanization and the Gentrification Process:

Projects involving transportation infrastructure can have a big influence on community cohesion, land use patterns, and urban growth. Preserving cheap housing close to transit hubs, keeping vulnerable people from being uprooted, and involving local communities in the planning and decision-making process are all examples of ethical considerations.

Health Public:

Decisions about transportation have an impact on public health outcomes, such as levels of physical activity, air quality, and road safety. Promoting environmentally friendly forms of transportation like walking and bicycling, lowering exposure to noise pollution and air pollution, and giving infrastructural support for sustainable and healthy lifestyles top priority are all examples of ethical concerns.

Inclusivity and Accessibility:

Transportation systems ought to be built with a variety of demographics in mind, such as the elderly, the disabled, the underprivileged, and marginalized communities. Ensuring barrier-free access to transportation options, offering accessible public transit services, and addressing transportation imbalances are all examples of ethical responsibilities.

Impacts on Culture and Society:

Social interactions, urban vibrancy, and cultural heritage can all be negatively impacted by changes in transportation technology and mobility patterns. Preserving community identity, maintaining social cohesiveness, and encouraging inclusive urban areas that suit a range of needs and preferences are all ethical considerations.

Design and Governance Ethics:

In order to advance sustainability, justice, and human well-being, ethical concepts must be considered while designing transportation systems and technology. Stakeholder engagement procedures, accountability systems, and ethical governance frameworks all aim to guarantee that transportation policies and practices respect moral principles and advance the general welfare.

Proactive policy interventions, stakeholder engagement, and interdisciplinary collaboration are necessary to address these ethical considerations and societal repercussions. Equitable, sustainable, and human-centered design principles can help us create transportation systems that improve people's quality of life, advance social justice, and benefit both the current and future generations.

Summary

To sum up, there is a great deal of promise for the future of technology to change almost every facet of human existence, including how we move, heal, and interact with our surroundings. As we keep expanding the frontiers of invention, a number of important themes become apparent:

Transformational Impact: New technologies with the potential to completely transform economies, industries, and communities around the world include biotechnology, artificial intelligence, renewable energy, space exploration, and the Internet of Things. These inventions will solve urgent global issues, advance progress, and increase efficiency.

Ethical Concerns: Privacy, security, equality, and environmental sustainability are among the ethical issues that these developments raise. In order to make sure that technology works for everyone in society and serves the common good, it is imperative that these ethical issues be addressed early on.

Collaborative Approach: Cooperation across disciplines, sectors, and geographies is necessary to address the complex challenges and opportunities posed by future technology. We can fully utilize technology to benefit humanity by promoting inclusive decision-making processes, interdisciplinary collaboration, and knowledge sharing.

The concept of responsible innovation pertains to the equitable integration of technology advancements with ethical, social, and environmental factors. It necessitates a dedication to openness, responsibility, and ongoing evaluation of the effects evolving technologies have on society.

Empowerment and Inclusion: By increasing access to economic opportunities, healthcare, education, and knowledge, future technologies have the potential to empower people, communities, and entire countries. Ensuring that technological innovations are inclusive and accessible to all is imperative in order to reduce inequities and bridge the digital divide.

Agility and Adaptability: In order to survive in a world that is changing quickly due to the rapid advancement of technology, both individuals and organizations need to develop their agility, adaptability, and capacity for lifelong learning. Navigating the complexity of the future will require adopting a growth mindset and cultivating an innovative culture.

In conclusion, technology has a lot of potential to advance human progress and bring about positive change in the future. Through responsible innovation adoption, collaborative efforts, and ethical concerns as top priorities, we may create a future in which technology acts as a positive force, improving the prosperity and well-being of people everywhere.

THE END